Student Book 1

Glenda Bradley

NELSON CENGAGE Learning

Australia • Brazil • Japan • Korea • Mexico • Singapore • Spain • United Kingdom • United States

Contents

Numbers! Numbers!

- Fill in the missing numbers.

0	I				5			9
II			14			18		

- How old will you be on your next birthday? _____

- What is the number on your classroom door? _____

- What is your favourite number? _____

  Why? _____

Play a game.

You will need: cards made from BLM I 'Number Tiles: 0–19' placed in a paper bag, a partner

- Write the numbers from 0 to 20 in any of the spaces below. You can repeat numbers.

- Play with a partner. In turn, take a number card from the bag. If you have that number on your number grid, cross it off.

- Place the number back in the bag.

- In turn, keep taking out cards until one player has crossed off all their numbers in their grid. This player is the winner.

Number Word Search

- Draw lines matching the words and the numerals.

1	four	11	nineteen
2	eight	12	fourteen
3	ten	13	seventeen
4	two	14	twelve
5	seven	15	fifteen
6	five	16	eighteen
7	nine	17	thirteen
8	one	18	twenty
9	six	19	eleven
10	three	20	sixteen

- Circle the number words for the numbers in the box.

1	11
2	12
3	13
4	14
5	15
6	16
7	17
8	18
9	19
10	20

s	e	v	e	n	t	e	e	n	k	t	t
l	f	i	v	e	h	o	s	s	n	e	w
e	o	s	i	x	i	t	e	p	i	n	e
i	u	f	o	u	r	w	v	e	n	o	l
g	r	i	y	b	t	e	e	l	e	e	v
h	t	f	m	a	e	n	n	e	t	l	e
t	e	t	t	w	e	t	l	v	e	t	i
e	e	e	w	i	n	y	e	e	e	h	g
e	n	e	o	n	e	z	s	n	n	r	h
n	i	n	e	q	r	y	w	e	p	e	t
t	x	i	v	n	s	i	x	t	e	e	n

Unit 1 **Recognising Numbers to 20** (TRB pp. 24–27)
Whole numbers MA1-4NA applies place value, informally, to count, order, read and represent two- and three-digit numbers

5

Classroom Things

- Draw 17 pencils in the tin.

  Colour some red, some blue and some green.

  How many pencils are in the tin? _____

  How many are red? _____

  How many are blue? _____

  How many are green? _____

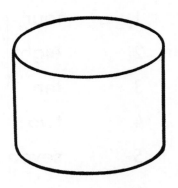

- Draw some books on the table.
  Some can be small
  and some can be big.

  How many books
  are on the table? _____

  How many are small? _____

  How many are big? _____

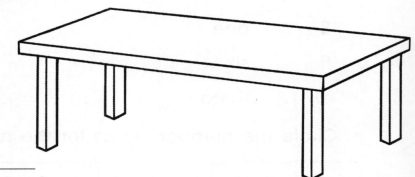

- Look for something else in the classroom.
  Draw your own picture to show 12.

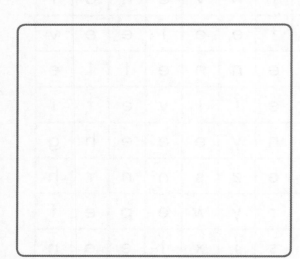

Write about your picture.

_____

_____

_____

_____

Unit 1 **Recognising Numbers to 20** (TRB pp. 24–27)
Whole numbers MA1-4NA applies place value, informally, to count, order, read and represent two- and three-digit numbers

STUDENT ASSESSMENT

- Write the numbers from 0 to 20.

  _____

  _____

- Write the number word for each numeral.

  12 _____   15 _____

  19 _____   13 _____

  7 _____   18 _____

  11 _____   14 _____

- Draw lines matching the words and the numerals.

  sixteen   20

  ten   16

  seventeen   8

  twenty   10

  eight   17

- Draw:

17	14
11	13

Unit
1
Recognising Numbers to 20 (TRB pp. 24–27)
Whole numbers MA1-4NA applies place value, informally, to count, order, read and represent two- and three-digit numbers

7

Counting Backwards

You will need: a calculator, a counter for each player, a dice, a partner

- Use a calculator to count backwards and help you fill in the spiral number board. This is what you do:

  * Press **20 – 1 =** and fill in the next number.

  * Press **=** and fill in the next number.

  * Keep pressing **=** and filling in the numbers to complete the spiral number board. Remember to follow the arrows.

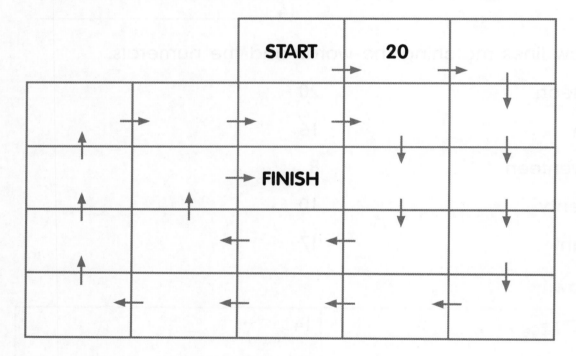

- Play a game with a partner.

  * Place your counter on START. Take turns to roll a dice and move your counter that many spaces.

  * Make sure you follow the counting backwards pattern.

  * The first player to reach FINISH wins.

Missing Numbers

- Write the number that comes **after**:

14 _____   7 _____   12 _____   16 _____

6 _____   19 _____   10 _____   13 _____

- Write the number that comes **before**:

_____ 12   _____ 5   _____ 10   _____ 17

_____ 19   _____ 8   _____ 13   _____ 20

- Fill in the missing numbers on each number line.

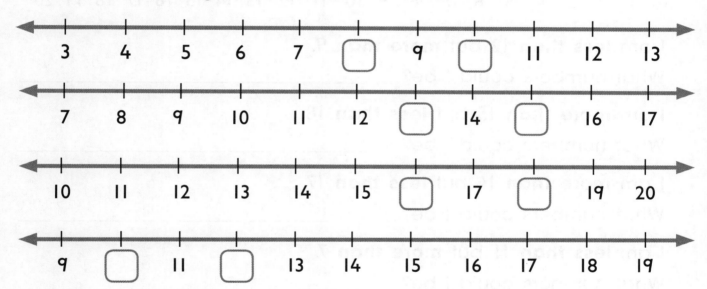

- What is my mystery number?

I come **before** 12 _____   I come **after** 13 _____

I come **after** 15 _____   I come **before** 20 _____

I come **before** 18 _____   I come **after** 9 _____

- Now it is your turn. Write a clue for a mystery number.

_____

Unit **2** **Counting to 20** (TRB pp. 28–31)
Whole numbers MA1-4NA applies place value, informally, to count, order, read and represent two- and three-digit numbers

9

Mystery Numbers

- Order the numbers from **smallest** to **largest**.

16	14	17	15				
18	14	16	12				
6	0	15	7				

9	11	10	12				
17	9	4	13				
11	19	16	13				

- Use the number line to help you find the mystery numbers.

0 1 2 3 4 5 6 7 8 9 10 11 12 13 14 15 16 17 18 19 20

I am **less than** 12 but **more than** 9.
What numbers could I be? _____

I am **more than** 15 but **less than** 18.
What numbers could I be? _____

I am **more than** 14 but **less than** 17.
What numbers could I be? _____

I am **less than** 11 but **more than** 7.
What numbers could I be? _____

I am **less than** 16 but **more than** 13.
What numbers could I be? _____

- Write your own clue for each number.

17 _____

8 _____

12 _____

2 STUDENT ASSESSMENT

DATE:

- Finish the counting sequences.

  0, 1, 2, 3, 5, _____, _____, _____, _____, _____, _____, _____, _____

  20, 19, 18, 17, _____, _____, _____, _____, _____, _____, _____, _____

- Write the number that comes **before**:

  _____ 7   _____ 18

  _____ 11   _____ 12

  _____ 16   _____ 20

  _____ 8   _____ 15

- Write the number that comes **after**:

  9 _____   17 _____

  11 _____   14 _____

  19 _____   10 _____

  13 _____   15 _____

- Order the numbers from **smallest** to **largest**.

  17 14 16 15   ▢ ▢ ▢ ▢

  19 3 14 7   ▢ ▢ ▢ ▢

  12 8 4 15   ▢ ▢ ▢ ▢

  20 11 6 17   ▢ ▢ ▢ ▢

Unit 2   **Counting to 20** (TRB pp. 28–31)
Whole numbers MA1-4NA applies place value, informally, to count, order, read and represent two- and three-digit numbers

11

Counting by 2s

• Write the counting pattern to work out how many feet altogether.

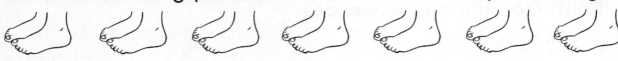

_____ _____ _____ _____ _____ _____ _____

There are _____ feet altogether.

• Write the counting pattern to work out how many mittens altogether.

_____ _____ _____ _____ _____

There are _____ mittens altogether.

• Write the counting pattern to work out how many eyes altogether.

_____ _____ _____ _____ _____ _____ _____

There are _____ eyes altogether.

• Write the counting pattern to work out how many socks altogether.

_____ _____ _____ _____

There are _____ socks altogether.

• Draw 6 pairs of shoes.

How many shoes
altogether? _____

Counting by 5s

- Finish writing the numbers in the table.

  Look at the numbers in the shaded column. Write them down.

  _____, _____, _____, _____,

  What is the pattern counting by? _____

1	2	3	4	5
6				

- Finish writing the numbers in the table. This time begin counting from 3.

  Look at the numbers in the shaded column. Write them down.

  _____, _____, _____, _____

  This pattern is also counting by 5s.

3	4	5		

- Choose your own number and start counting from it.

  Look at the numbers in the shaded column. Write them down.

  _____, _____, _____, _____

  What is this pattern counting by? _____

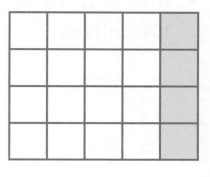

  What do you think will be the next number in the pattern? _____

  Explain how you worked it out. _____

  _____

Unit 3 **Skip Counting** (TRB pp. 32–35)
Multiplication and division MA1-6NA uses a range of mental strategies and concrete materials for multiplication and division

13

Skip Counting

Look at the 100 chart below.

- Colour 2 in yellow. Keep counting by 2s and colouring each number yellow.

  What is the pattern? _____

- Colour 5 in red. Keep counting by 5s and colouring each number red.

  What is the pattern? _____

- Write only the numbers you coloured that were in both the 2s and 5s pattern.

  _____

- Colour 10 in blue. Keep counting by 10s and colouring each number blue.

  What is the pattern? _____

- Write only the numbers you coloured that were in the 2s, 5s and 10s pattern.

  _____

  _____

1	2	3	4	5	6	7	8	9	10
11	12	13	14	15	16	17	18	19	20
21	22	23	24	25	26	27	28	29	30
31	32	33	34	35	36	37	38	39	40
41	42	43	44	45	46	47	48	49	50
51	52	53	54	55	56	57	58	59	60
61	62	63	64	65	66	67	68	69	70
71	72	73	74	75	76	77	78	79	80
81	82	83	84	85	86	87	88	89	90
91	92	93	94	95	96	97	98	99	100

STUDENT ASSESSMENT

- Finish the counting patterns.

    0, 2, 4, _____, _____, _____, _____, _____, _____

    6, 8, 10, _____, _____, _____, _____, _____, _____

- Write the missing numbers.

    0, 2, 4, _____, _____, _____, 12, 14, 16, _____, _____

- Finish the counting pattern.

    0, 5, 10, _____, _____

- Finish the counting pattern.

    0, 10, 20, _____, _____

- Write the missing numbers.

    20, 25, 30, _____, _____, 45

- Write the missing numbers.

    20, 30, 40, _____, _____, 70

- I need to know how many 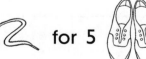 for 5 👞 .

    What counting pattern will help me work it out? _____

Unit
3
Skip Counting (TRB pp. 32–35)
Multiplication and division MA1-6NA uses a range of mental strategies and concrete materials for multiplication and division

15

In the Classroom

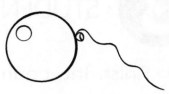

- Find and draw **5** objects in your classroom that are the shape of a:

rectangle

square

circle

- What other interesting shapes can you see in the classroom? Draw and name them below.

2D Shapes (TRB pp. 36–39)
Two-dimensional space MA1-15MG manipulates, sorts, represents, describes and explores two-dimensional shapes, including quadrilaterals, pentagons, hexagons and octagons

What's the Difference?

- Colour the shapes with ☺ blue and shapes with ♥ yellow.

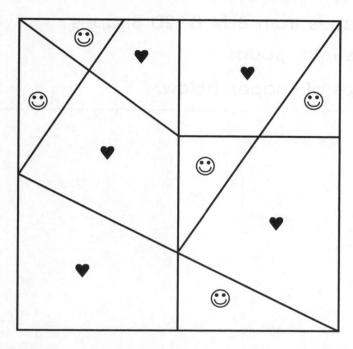

- What is the difference between the blue and the yellow shapes?

_____

- Draw some other shapes that could belong to the group of yellow shapes.

Unit 4 **2D Shapes** (TRB pp. 36–39)
Two-dimensional space MA1-15MG manipulates, sorts, represents, describes and explores two-dimensional shapes, including quadrilaterals, pentagons, hexagons and octagons

17

Sorting Shapes

You will need: a copy of BLM 8 '2D Shapes', scissors, glue

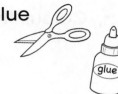

- Cut out the shapes from BLM 8 '2D Shapes'.
  Sort the shapes into groups.
- Paste the groups of shapes below.

- Explain how you grouped your shapes.

_____

- Could you have grouped them another way? Explain.

_____

STUDENT ASSESSMENT

• Draw a:

square	circle	triangle
rectangle	hexagon	

• Draw and name any other shapes you know.

• Draw a shape that has 3 corners.

• Draw two shapes that have 4 edges.

• Draw another shape that belongs to each group.

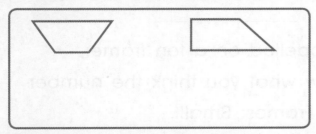

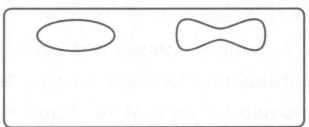

Unit
4

2D Shapes (TRB pp. 36–39)
Two-dimensional space MA1-15MG manipulates, sorts, represents, describes and explores two-dimensional shapes, including quadrilaterals, pentagons, hexagons and octagons

19

Numbers on a Ten Frame

DATE:

You will need: a copy of BLM 10 'Blank Ten Frames: Small'

- Show each number on the ten frames.

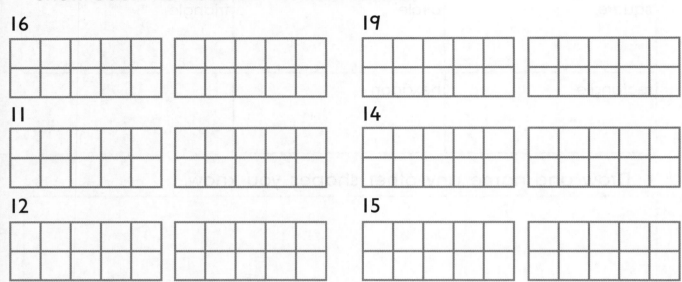

16

19

11

14

12

15

- Look at the ten frames. Write the numbers you can see.

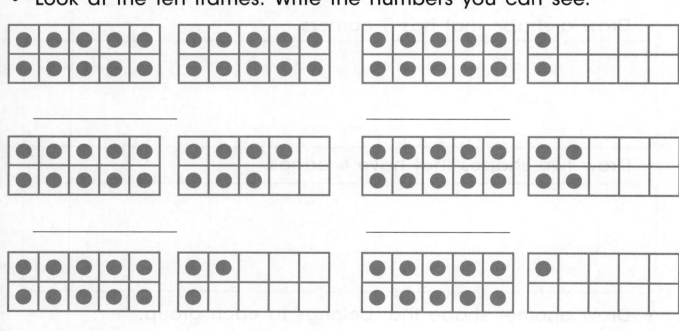

- A mystery number has been modelled on a ten frame.
  It has one ten and **?** ones. Draw what you think the number
  could be on BLM 10 'Blank Ten Frames: Small'.

Unit **5** **Modelling Numbers** (TRB pp. 40–43)
Whole numbers MA1-4NA applies place value, informally, to count, order, read and represent two- and three-digit numbers

Colouring Cubes

You will need: a dice, a partner

• You will roll the dice **5** times. Each time you roll, you will colour that many cubes. Before you begin, answer this question:

How many cubes do you think you will colour altogether? _____

• Roll the dice once and colour that many cubes.

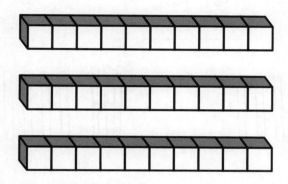

• Roll the dice 4 more times and colour the cubes for each roll.

How many bars of 10 did you colour in? _____

How many cubes did you colour altogether? _____

• Ask your partner their score and write it down. _____

How many bars of 10 will you need to colour for their score? _____

Show your partner's score on the cubes below.

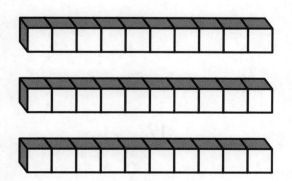

Unit 5 **Modelling Numbers** (TRB pp. 40–43)
Whole numbers MA1-4NA applies place value, informally, to count, order, read and represent two- and three-digit numbers

21

Bundles and Sticks

You will need: a copy of BLM 11 'Craft Sticks and Bundles', scissors, glue

- Write the number.

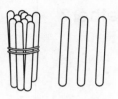   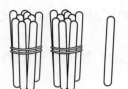   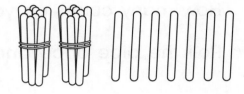

_____   _____   _____

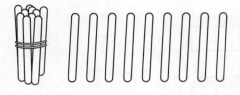   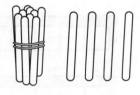

_____   _____

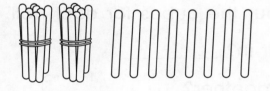   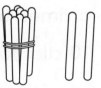

- Model each number using BLM 11 'Craft Sticks and Bundles'. Paste your model below.

22   17   26

5 STUDENT ASSESSMENT

- Show the numbers on the ten frames.

10

18

12

16

- Write the numbers.

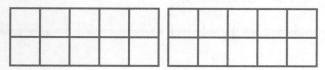

_____

_____

_____

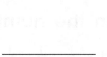 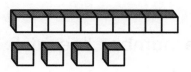

_____

- Draw lines matching each amount.

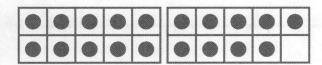

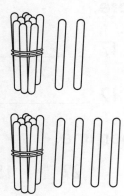

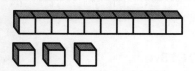

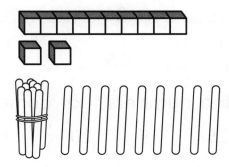

Unit
5
Modelling Numbers (TRB pp. 40–43)
Whole numbers MA1-4NA applies place value, informally, to count, order, read and represent two- and three-digit numbers

23

Number Lines

You will need: If you need help, see NTO 1.1 'Number Tiles'

- Finish the number line.

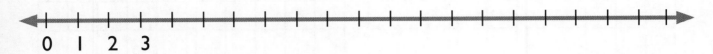

0 1 2 3

- Start the number line at 6. Fill in the numbers.

- Start the number line at 3. Fill in the numbers.

- Start the number line at 9. Fill in the numbers.

- Start the number line at 5. Fill in the numbers.

- Start the number line at 11. Fill in the numbers.

- On a number line, what number would come **before**:

22 _____   26 _____   20 _____   17 _____

30 _____   29 _____   15 _____   12 _____

- Circle the numbers that come after 18 on a number line.

17   25   19   20   13   21   16   11   27   15

Numbers on the Line

DATE:

- Fill in the missing numbers.

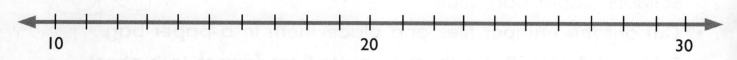

10   20   30

- Write three numbers that are **more than** 21. _____

- Write three numbers that are **less than** 17. _____

- Write three numbers that are **more than** 17
  but **less than** 21. _____

- Write three numbers that are **more than** 24. _____

- Write three numbers that are **less than** 12. _____

- Write three numbers that are **more than** 12
  but **less than** 24. _____

- Write a clue for each number. Read the clue to your partner.
  Have them guess your number.

14 _____

27 _____

18 _____

23 _____

Unit
6
Number Lines (TRB pp. 44–47)
Whole numbers MA1-4NA applies place value, informally, to count, order, read and represent two- and three-digit numbers

25

Order the Numbers

You will need: a copy of BLM 1 'Number Cards: 0–19', scissors, paper bag, glue

- Cut out the number tiles and place them in a paper bag. Take out **3** tiles. Paste them in order from **lowest** to **highest**.

- Select **4** more cards. Paste them in order from **highest** to **lowest**.

- Select **5** more cards. Paste them in order from **lowest** to **highest**.

- After a game, the scores were:
  Lee 21, Jill 18, Sam 12, Iman 25, Sasha 28, Ravneet 19
  Who won the game? _____
  Order the scores from **highest** to **lowest**.

STUDENT ASSESSMENT

DATE:

- Look at the number line.

0   1   2   6   11   12   16   17   18

What numbers are missing? _____

_____

What numbers are **less than** 9? _____

_____

What numbers are **more than** 18? _____

_____

What numbers are **more than** 11 but **less than** 17? _____

_____

- Order the numbers from **lowest** to **highest**.

25   13   19   21   12

_____

- Order the numbers from **highest** to **lowest**.

11   17   23   12   22

_____

Unit 6   **Number Lines** (TRB pp. 44–47)
Whole numbers MA1-4NA applies place value, informally, to count, order, read and represent two- and three-digit numbers

27

My Bedroom

- Draw a  **next to** the teddy.

- Draw a mouse **under** the wardrobe.

- Draw some  **on** the bed.

- Draw some books **in front of** the chair.

- Draw a tree **outside** the door.

- Draw a toy car **between** the bed and the teddy.

- Draw some books **in** the wardrobe.

- Draw a ball **behind** the teddy.

- Draw other things in your picture. On another sheet of paper, write the instructions.

Which Way?

- Draw a path from START to HOME.

  You cannot go through the square with the house or the pond.

  Try to go through all the other squares only once.

START			
			HOME

- If you were walking along the path, would you have made any turns? Put a red circle around the **clockwise turns** and a blue circle around the **anticlockwise turns**.

- Can you find another path? Use another coloured pencil to draw it.

Unit **7** **Position** (TRB pp. 48–51)
Position MA1-16MG represents and describes the positions of objects in everyday situations and on maps

29

Follow the Directions

You will need: a partner

- With your partner, take it in turns to describe **one** of the paths you drew in 'Which Way?' on page 29.

   If your directions are clear, your partner's map will match yours. Check when you finish.

START			
			HOME

- List some location words you used.

_____

_____

DATE:

STUDENT ASSESSMENT

• Write each word under its matching picture.

inside   near   anticlockwise   under   between   clockwise   behind

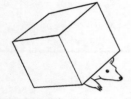         

_____   _____   _____   _____

_____   _____

• Colour this path on the picture.

✻ Start walking from the path **near** the vegetable garden. Walk **past** the cows.

✻ Turn **clockwise** and follow the path **between** the horses and cows until you meet another path.

✻ Turn **anticlockwise** and move **past** the horses until you meet another path.

✻ Turn **anticlockwise** and move **between** the sheep and the horses. Follow the path to where you started.

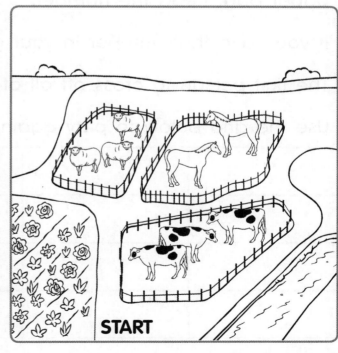

START

Unit
7   **Position** (TRB pp. 48–51)
Position MA1-16MG represents and describes the positions of objects in everyday situations and on maps

31

Number Bingo

You will need: a partner, a copy of BLM 15 '100 Chart', scissors, a paper bag

- Choose numbers between 1 and 100.

  Write each number in the grid below.

- Cut up BLM 15 '100 Chart' to make number tiles.

  Place the number tiles in a paper bag.

- With your partner, take it in turns to pull a number tile from the paper bag. Read the number.

- If you have that number in your grid, cross it off.

- The first person to cross off all of their numbers wins.

- Use the grid below to play again.

Lucky Slides

You will need: a dice, small counters

Fill in the missing **numbers** to complete the game board.

0	1	2		4	5			8	9
10		13	14	15	16	17			19
	21	22		24			27		
30		33		35				38	39
40				45	46	47			
	51	52		55		57	58		
60		63	64		66	67	68		
	71	72			76		78	79	
80			84	85	86				
90	91					97	98	99	

- To play a game with a partner, put your counters on 0.

- In turn, roll the dice and move that many spaces.

- Look carefully so you follow the numbers in the correct order. If you do not follow the correct order, you have to go back to 0.

- If you land on a number at the top of the slide, "slide" down to the number at the bottom.

- The first to 99 wins!

Unit 8 **Numbers Beyond 20** (TRB pp. 52–55)
Whole numbers MA1-4NA applies place value, informally, to count, order, read and represent two- and three-digit numbers

33

What's the Number?

You will need: a set of number tiles made from
BLM 15 '100 Chart' for each table, glue

* Select a number tile. Paste it in the space below. Think about a
100 chart. Now write the number that comes **before** and **after**.
Do this 3 more times.

_____ ☐ _____   _____ ☐ _____

_____ ☐ _____   _____ ☐ _____

* Three groups of children played a game.

Write who came 1st, 2nd and 3rd for each group.

Their scores were:

  * *Abdul: 78,*   *Jake: 79,*   *Lee: 76*

    1st _____   2nd _____   and 3rd _____

  * *Lucy: 84,*   *Iman: 86,*   *Maria: 85*

    1st _____   2nd _____   and 3rd _____

  * *Carlos: 39,*   *Nat: 41,*   *Tom: 40*

    1st _____   2nd _____   and 3rd _____

* Jack scored 79, Jan scored **one more** than Jack, Jo scored
one less than Jack and Jed scored **one more** than Jan.

What were their scores? Write them in order.

_____

STUDENT ASSESSMENT

You will need: a set of number tiles made from BLM 15 '100 Chart'

- Have your teacher or a partner select 8 number tiles. As they read them to you, write down each number.

_____

- Fill in the missing numbers in the 100 chart.

1									
								19	
			24						
									50
		53							
				75					
									100

- Write the numbers that come **after** each number.

71 _____   17 _____

39 _____   63 _____

Numbers Beyond 20 (TRB pp. 52–55)
Whole numbers MA1-4NA applies place value, informally, to count, order, read and represent two- and three-digit numbers

Show the Time

- Show the times on each clock.

 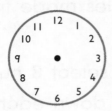 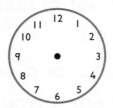 

6 o'clock   3 o'clock   12 o'clock   half-past 9

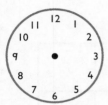  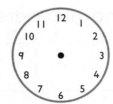 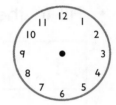

half-past 3   half-past 7   10 o'clock   half-past 1

- Write the time shown on each clock.

_____   _____   _____   _____

_____   _____   _____   _____

- Show your favourite time of the day on this clock.
  Write why it is your favourite time.

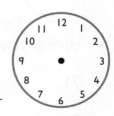

_____

_____

Digital Times

- Look at the times on each clock.

- Write the matching digital time. One has been done.

4:30

_____   _____   _____   _____

_____   _____   _____   _____

_____   _____   _____   _____

- List the times above in order. Begin with 2:00.

2:00 , _____

- Write what you could be doing at the **earliest** time?

_____

- Write what you could be doing at the **latest** time?

_____

Unit 9
Time (TRB pp. 56–59)
Time MA1-13MG describes, compares and orders durations of events, and reads half- and quarter-hour time

37

Your Day

• Show the time on each clock. Draw or write what you might do at those times.

8:30	
1 o'clock	
3:00	
half-past 6	

Unit 9 **Time** (TRB pp. 56–59)
Time MA1-13MG describes, compares and orders durations of events, and reads half- and quarter-hour time

STUDENT ASSESSMENT

• **Complete the table.**

half-past 7	(clock face)	▢▢ : ▢▢
	(clock face)	5 : 0 0
	(clock showing hands near 12 and 1)	▢▢ : ▢
	(clock showing hands near 12 and 6)	▢▢ : ▢▢
	(clock face)	3 : 3 0

• **Circle the clock that shows your favourite time of the day.**

• **Explain why it is your favourite time.**

_____

_____

_____

Unit
9
Time (TRB pp. 56–59)
Time MA1-13MG describes, compares and orders durations of events, and reads half- and quarter-hour time

39

Two Footprints Long

You will need: a copy of BLM 19 'Dinosaur Footprints', scissors

- Cut out your dinosaur footprints.

- Look around your classroom. Draw 3 objects in the classroom that might be 2 footprints long.

- Now measure the objects and record what you found out.

Objects (Draw here)	How many footprints long?	Write what you found out when you measured.

- How did you decide which objects might be long?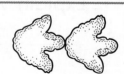

_____

_____

 Length and Area (TRB pp. 60–63)
Length MA1-9MG measures, records, compares and estimates lengths and distances using uniform informal units, metres and centimetres

Paperclips or Counters?

You will need: paperclips, counters

- Choose what you are going to use to measure the length of each object. Write it here:

_____

- First estimate how long you think each object will be.

- Now measure each object.

Object	Estimate	How many?
Book		
Straps on school bag		

- Which object is longer?

_____

How do you know?

_____

- Check with a partner. Did they get the same results as you?

_____

How do you think that happened?

_____

Unit 10 **Length and Area** (TRB pp. 60–63)
Length MA1-9MG measures, records, compares and estimates lengths and distances using uniform informal units, metres and centimetres

41

Area

You will need: counters

- Use counters to find the area of each shape.
  How many counters for each shape?

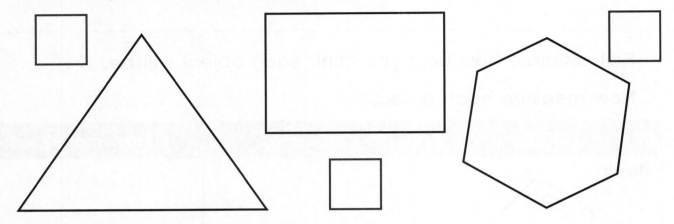

  Colour the shape with the largest area red.
  Colour the shape with the smallest area blue.
  Is it hard to cover the triangle and hexagon with counters? _____

- Count the squares to find the area of each letter.

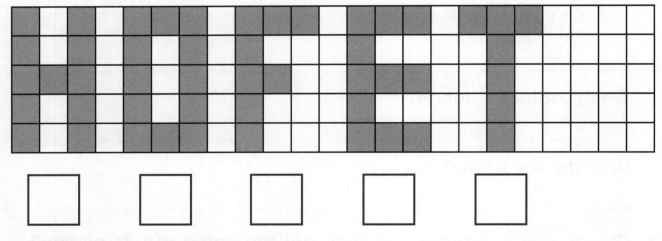

  Which has the **largest** area? _____
  Which has an area that is 1 square **bigger than** 'T'? _____
  Which have the **same** area? _____
  Draw a letter 'L' that has an area of 7 squares.

STUDENT ASSESSMENT

DATE:

You will need: counters

- Look at the snakes. Estimate how long you think each snake might be if you measured it with counters.

- Now measure the snakes with counters.

	Estimate	Number of counters long
Sammy		
Sid		

- Which two letters have the same area? Use counters to find out.

The letters with the same area are _____ and _____.

Unit
10
Length and Area (TRB pp. 60–63)
Length MA1-9MG measures, records, compares and estimates lengths and distances using uniform informal units, metres and centimetres
Area MA1-10MG measures, records, compares and estimates areas using uniform informal units

43

Toys, Toys, Toys

- Write how many of each toy. Work out how many altogether.

_____ + _____ = _____   _____ + _____ = _____

 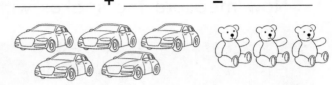

_____ + _____ = _____   _____ + _____ = _____

_____ + _____ = _____   _____ + _____ = _____

_____ + _____ = _____   _____ + _____ = _____

- There were 7 toys altogether. Draw a
  picture of what they could be and write
  a number sentence for your drawing.

_____ + _____ = _____

What other numbers joined together make 7?

_____ + _____ = _____   _____ + _____ = _____

Unit 11 **Addition** (TRB pp. 64–67)
Addition and Subtraction MA1-5NA uses a range of strategies and informal recording methods for addition and subtraction
involving one- and two-digit numbers

Count On

• The number on the toy box tells us how many toys are in the box.
Count on to work out how many toys altogether.

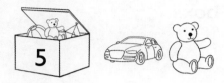

5 + 2 = _____

9 + 4 = _____

6 + 4 = _____

4 + 5 = _____

8 + 3 = _____

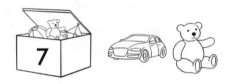

7 + 2 = _____

6 + 4 = _____

8 + 6 = _____

• Make your own number story and complete the number sentence.

_____ + _____ = _____   _____ + _____ = _____

Unit **11** **Addition** (TRB pp. 64–67)
Addition and Subtraction MA1-5NA uses a range of strategies and informal recording methods for addition and subtraction involving one- and two-digit numbers

45

Counting On 1, 2 or 3

You will need: a dice

- Roll the dice and write the number in the box.

  Count on to find the total.

  $$\begin{array}{r} \boxed{} \\ +\ \ 2 \\ \hline \end{array}$$

- Repeat and count on to find the total for each problem.

$$\begin{array}{r} \boxed{} \\ +\ \ 3 \\ \hline \\ \hline \end{array} \qquad \begin{array}{r} \boxed{} \\ +\ \ 1 \\ \hline \\ \hline \end{array} \qquad \begin{array}{r} \boxed{} \\ +\ \ 2 \\ \hline \\ \hline \end{array} \qquad \begin{array}{r} \boxed{} \\ +\ \ 3 \\ \hline \\ \hline \end{array}$$

$$\begin{array}{r} \boxed{} \\ +\ \ 1 \\ \hline \\ \hline \end{array} \qquad \begin{array}{r} \boxed{} \\ +\ \ 2 \\ \hline \\ \hline \end{array} \qquad \begin{array}{r} \boxed{} \\ +\ \ 3 \\ \hline \\ \hline \end{array} \qquad \begin{array}{r} \boxed{} \\ +\ \ 2 \\ \hline \\ \hline \end{array}$$

$$\begin{array}{r} \boxed{} \\ +\ \ 2 \\ \hline \\ \hline \end{array} \qquad \begin{array}{r} \boxed{} \\ +\ \ 3 \\ \hline \\ \hline \end{array} \qquad \begin{array}{r} \boxed{} \\ +\ \ 1 \\ \hline \\ \hline \end{array} \qquad \begin{array}{r} \boxed{} \\ +\ \ 3 \\ \hline \\ \hline \end{array}$$

STUDENT ASSESSMENT

- Write how many dots on each side of the ladybird.
Then work out how many dots altogether.

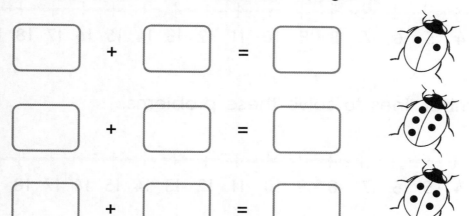

- There are 6 insects under the leaf, count on to work out how many altogether.

6 + 3 = _____

- There are 7 insects under the leaf, count on to work out how many altogether.

7 + 1 = _____

- There are 5 insects under the leaf, count on to work out how many altogether.

5 + 2 = _____

- Count on to find the total.

Unit
11
Addition (TRB pp. 64–67)
Addition and Subtraction MA1-5NA uses a range of strategies and informal recording methods for addition and subtraction involving one- and two-digit numbers

47

Number Lines for Addition

- This number line helps us to find the answer to $9 + 5 =$ _____

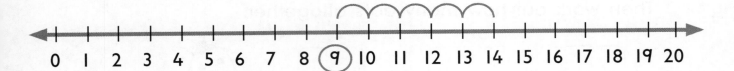

- Use the number lines to solve these problems.

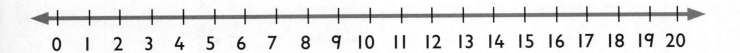

$5 + 3 =$ _____ $9 + 4 =$ _____ $16 + 3 =$ _____

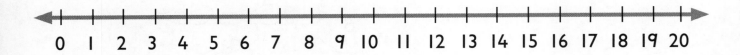

$2 + 9 =$ _____ $13 + 2 =$ _____ $15 + 5 =$ _____

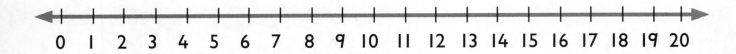

$3 + 8 =$ _____ $12 + 4 =$ _____ $13 + 5 =$ _____

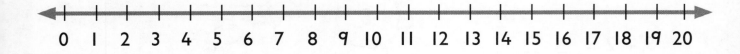

$12 + 4 =$ _____ $9 + 6 =$ _____ $7 + 2 =$ _____

 Developing Mental Strategies for Addition (TRB pp. 68–71)
Addition and subtraction MA1-5NA uses a range of strategies and informal recording methods for addition and subtraction involving one- and two-digit numbers

Tens Fact Search

- Circle the pairs of numbers that make 10. One has been done.

8	3	9	1	6
5	2	0	5	4
4	7	10	5	2
3	6	1	4	8
10	0	9	3	7

- Draw a circle around the dominoes that make 10.

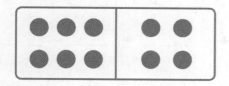
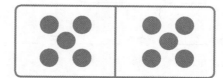

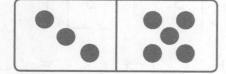
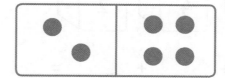

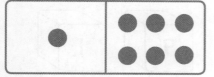
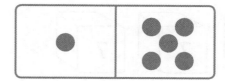

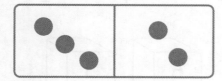
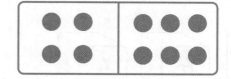

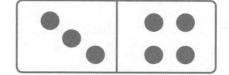

Unit 12 **Developing Mental Strategies for Addition** (TRB pp. 68–71)
Addition and subtraction MA1-5NA uses a range of strategies and informal recording methods for addition and subtraction involving one- and two-digit numbers

49

Make 10 and Add

- Circle the two numbers that make 10. Rearrange and add on the third number.

One has been done.

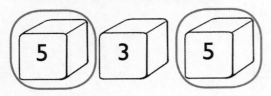   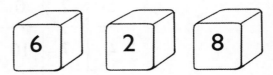

5 + 5 + 3 = 13   ___ + ___ + ___ = ___

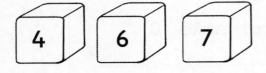   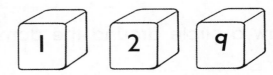

___ + ___ + ___ = ___   ___ + ___ + ___ = ___

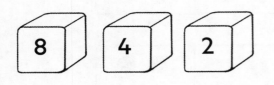   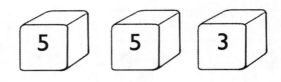

___ + ___ + ___ = ___   ___ + ___ + ___ = ___

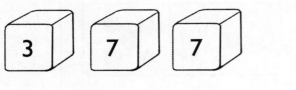   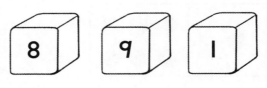

___ + ___ + ___ = ___   ___ + ___ + ___ = ___

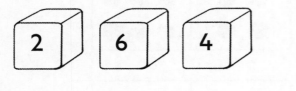   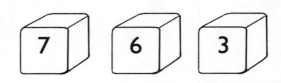

___ + ___ + ___ = ___   ___ + ___ + ___ = ___

DATE:

STUDENT ASSESSMENT

- On the number line show how you would solve these problems. Write your answer.

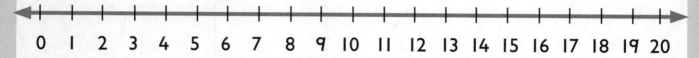

0 1 2 3 4 5 6 7 8 9 10 11 12 13 14 15 16 17 18 19 20

$3 + 4 =$ _____ $8 + 5 =$ _____ $17 + 2 =$ _____

- Write the number you need to add to make 10.

$6 +$ _____ $= 10$ $5 +$ _____ $= 10$

$7 +$ _____ $= 10$ $2 +$ _____ $= 10$

- Complete these problems.

$10 + 8 =$ _____ $10 + 5 =$ _____

$10 + 1 =$ _____ $10 + 3 =$ _____

$10 + 7 =$ _____ $10 + 6 =$ _____

- Find the total of these numbers.

           Total: _____

Which numbers did you add together first? _____

How did you add them together?

_____

How did you add the final number?

_____

Unit 12 **Developing Mental Strategies for Addition** (TRB pp. 68–71)
Addition and subtraction MA1-5NA uses a range of strategies and informal recording methods for addition and subtraction involving one- and two-digit numbers

51

Halves and Quarters

- Colour a half of each shape.

 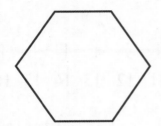 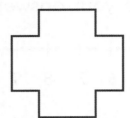 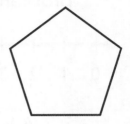

- Colour a quarter of each shape.

 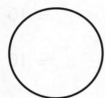 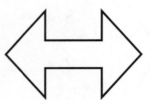 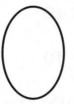  

- Look at the shaded part. Write $\frac{1}{2}$ or $\frac{1}{4}$ below each shape.

    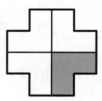

- Tell someone how you know that a quarter of this shape has been shaded.

Biscuits on the Plate

- Write and draw your answers.

There were 8 biscuits in the packet.
Draw half on each plate.

$\frac{1}{2}$ of 8 is _____

There were 12 biscuits in the packet.
Draw half on each plate.

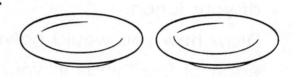

$\frac{1}{2}$ of 12 is _____

There were 4 biscuits in the packet.
Draw half on each plate.

$\frac{1}{2}$ of 4 is _____

There were 10 biscuits in the packet.
Draw half on each plate.

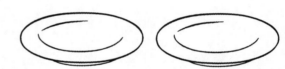

$\frac{1}{2}$ of 10 is _____

There were 6 biscuits in the packet.
Draw half on each plate.

$\frac{1}{2}$ of 6 is _____

Challenge!

Less than half the class were girls.
There were 9 girls in the class,
how many boys could there be?
Show how you worked out
your answer.

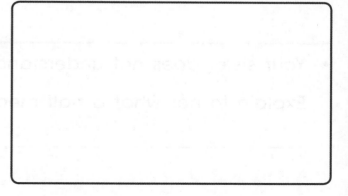

Sharing Lunch

- Imagine this is a picture of your lunch.

  Your sister has forgotten her lunch.

  You have to give her half of your lunch.

  Draw how you would halve each of the things in your lunchbox.

- Your sister does not understand what a half means.

  Explain to her what a half means. _____

  _____

  _____

STUDENT ASSESSMENT

- Is the shaded part $\frac{1}{2}$ or $\frac{1}{4}$?

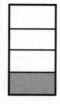

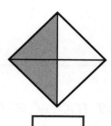

- Colour half of each shape.

- Colour a quarter of each shape.

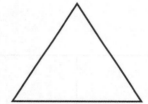
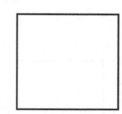
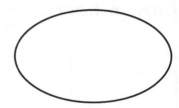

- Colour half and complete the number sentence.

$\frac{1}{2}$ of 6 is _____

$\frac{1}{2}$ of 10 is _____

$\frac{1}{2}$ of 4 is _____

$\frac{1}{2}$ of 8 is _____

Unit
13
Halves and Quarters (TRB pp. 72–75)
Fractions and decimals MA1-7NA represents and models halves, quarters and eighths

55

Making Groups

• Draw 5 smiley faces.

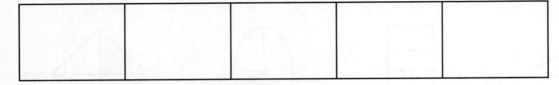

How many eyes? ☐

☐ groups of ☐ = ☐

• Draw 6 hands.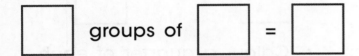

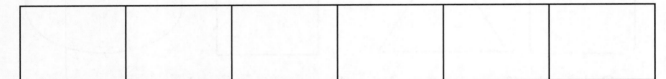

How many fingers? ☐

☐ groups of ☐ = ☐

• Draw 10 counters in each bowl.

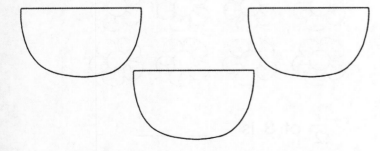

How many counters? ☐

☐ groups of ☐ = ☐

Equal Groups

• How many groups of 2?

 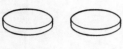

[ ] groups of 2 = [ ]   [ ] groups of 2 = [ ]

• How many groups of 5?

 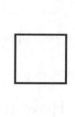

[ ] groups of 5 = [ ]   [ ] groups of 5 = [ ]

• How many groups of 10?

[ ] groups of 10 = [ ]

Sharing

- Share the pencils between the 2 tins.

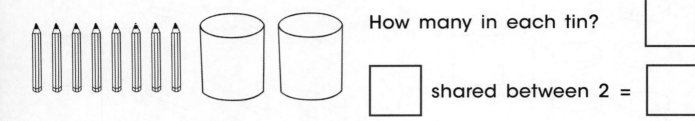

How many in each tin? ☐

☐ shared between 2 = ☐

- Share the grapes between the 3 plates.

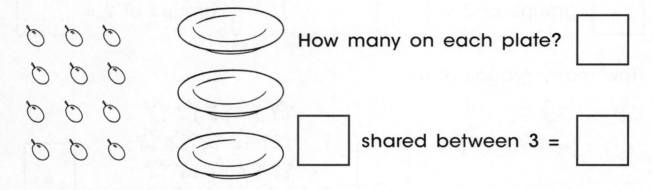

How many on each plate? ☐

☐ shared between 3 = ☐

- How many children can have 2 strawberries each?

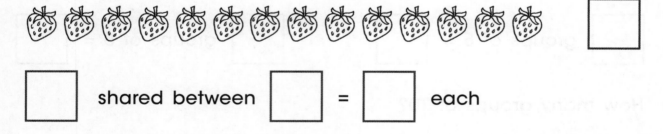

☐ shared between ☐ = ☐ each

- How many children can have 5 counters each?

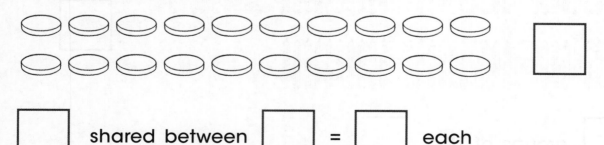

☐ shared between ☐ = ☐ each

Grouping and Sharing (TRB pp. 76–79)
Multiplication and division MA1-6NA uses a range of mental strategies and concrete materials for multiplication and division

DATE:

STUDENT ASSESSMENT

- How many groups of 2?

  ☐ groups of 2 = ☐

- How many groups of 10?

  ☐

  ☐ groups of 10 = ☐

- Draw three feet.

  How many toes? ☐

  ☐ groups of ☐ = ☐

- Share the grapes between 2 hands.

- How many in each hand? ☐

  ☐ shared between ☐ = ☐ each.

Unit **14** **Grouping and Sharing** (TRB pp. 76–79)
Multiplication and division MA1-6NA uses a range of mental strategies and concrete materials for multiplication and division

59

Roll Ten Times

You will need: a 10-sided dice, a partner

Play a game

- In turn, roll the dice. Draw that many dots in the first ten frame.

- Keep rolling the dice. When you have filled one ten frame, start drawing dots in the second ten frame.

- Check your total score after, 3, 6 and 10 rolls of the dice and fill in the table below.

Scores

	You	Partner
After 3 turns		
After 6 turns		
Final (after 10 turns)		

Craft Sticks

You will need: a copy of BLM 11 'Craft Sticks and Bundles', scissors, glue

- Write the number of craft sticks.

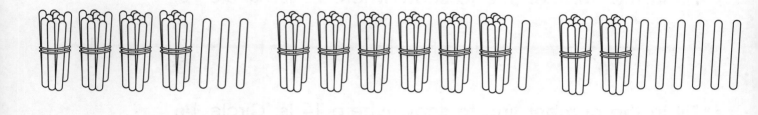

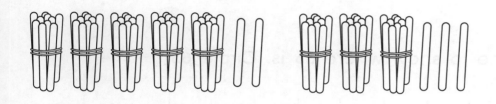

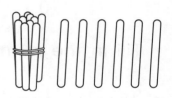

- Using BLM 11 'Craft Sticks and Bundles', paste the matching bundles and sticks beside each number.

27   61

- Using BLM 11 'Craft Sticks and Bundles', show:

  * how old your
    mother or father is.

  * how old you think
    your teacher is.

Unit **15** **2-digit Numbers** (TRB pp. 80–83)
Whole numbers MA1-4NA applies place value, informally, to count, order, read and represent two- and three-digit numbers

61

On the Number Line

- Fill in the number line to show where 38 is. Circle 38.

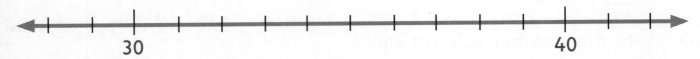

- Fill in the number line to show where 73 is. Circle 73.

- Fill in the number line to show where 14 is. Circle 14.

- Fill in the number line to show where 65 is. Circle 65.

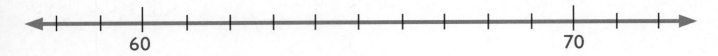

- Fill in the number line to show where 92 is. Circle 92.

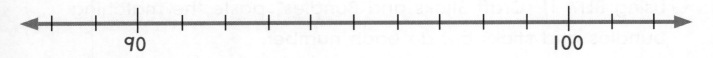

- If the number lines could be joined together, which circled number would be **furthest** along the number line? _____

How do you know? _____

_____

Which circled number is the **lowest** number? _____

How do you know? _____

STUDENT ASSESSMENT

- Write the number.

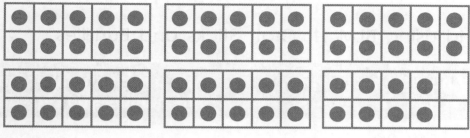

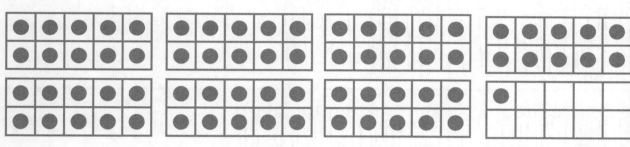

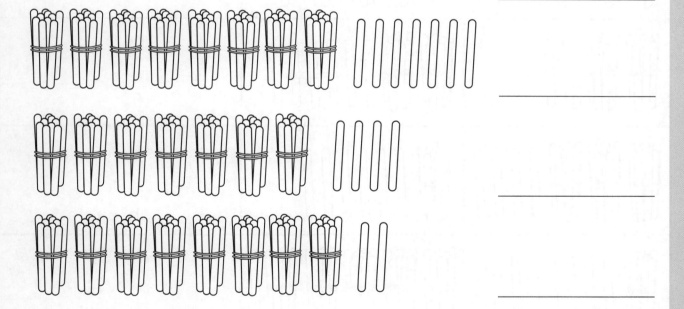

- Look at the numbers above. Are there any numbers that you can write on the number line? Write them on.

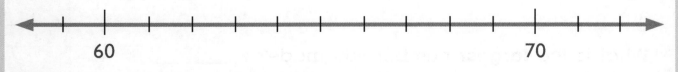

60   70

Unit
15
2-digit Numbers (TRB pp. 80–83)
Whiole numbers MA1-4NA applies place value, informally, to count, order, read and represent two- and three-digit numbers

63

Which Is Larger?

You will need: a copy of BLM II 'Craft Sticks and Bundles', scissors, glue

- Complete the grid using BLM II 'Craft Sticks and Bundles'.

Circle the largest number in each pair.		Paste a larger number than the number you circled.

- Write the numbers you modelled as numerals.

_____

What is the **largest** number you made? _____

Number Lines Can Help

- Circle the **smallest** number below.

    24   32   42   23   27

- Fill in the number line. Include the numbers from the first task. Begin with the **smallest** number.

    Circle the other numbers from the first task.

Write the numbers you circled from **largest** to **smallest**.

_____

- Order the numbers from **largest** to **smallest**. The number line can help you.

    41   27   39   33   31

_____

    26   20   33   38   29

_____

- Imagine you could keep counting on your number line. Order the numbers from **smallest** to **largest**.

    45   38   42   17   48

_____

    16   33   46   36   26

_____

Unit **16** **More About 2-digit Numbers** (TRB pp. 84–87)
Whole numbers MA1-4NA applies place value, informally, to count, order, read and represent two- and three-digit numbers

65

Largest Number Wins

You will need: two 10-sided dice (0–9), a partner

- Roll the two dice. Make the **largest** number you can and write it below.

- Write down your partner's number. Circle the **largest** number.

- The player with the **largest** number each round scores 1 point. The player with the most points after 10 turns wins.

	My number	My partner's number	I score	My partner scores
1				
2				
3				
4				
5				
6				
7				
8				
9				
10				
		Total		

- What was the **largest** number you made? _____

- What was the **smallest** number you made? _____

DATE:

STUDENT ASSESSMENT

You will need: a copy of BLM 11 'Craft Sticks and Bundles', scissors, glue

• Write the number for each model.

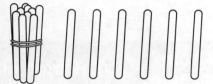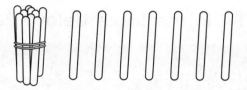

_____   _____

• Model each number using BLM 11 'Craft sticks and Bundles'.

          35   53

• Write the numbers on the number line.

    51   48   62   57

• Order the numbers from **smallest** to **largest**.

  27   47   19   38   92   4

_____

16 **More About 2-digit Numbers** (TRB pp. 84–87)
Whiole numbers MA1-4NA applies place value, informally, to count, order, read and represent two- and three-digit numbers

67

Different Views

You will need: 3D objects made from BLM 27 '3D Objects 1' and BLM 28 '3D Objects 2', a partner, sheet of paper

- Place a 3D object between you and your partner. Move around your object. Draw the view from two different sides and above. Fill in the table below. Repeat.

Name of object	View from side	View from another side	View from above

- Which 2D shapes did you see the most when looking at the faces of each 3D object?

  _____

- Could you work out a 3D object if you were only given one view of it? _____ Explain. _____

  _____

- On another sheet of paper, list the 3D objects you see most often in your classroom.

Corners, Edges and Faces

You will need: a set of 3D objects made from BLM 27 '3D Objects 1' and BLM 28 '3D Objects 2', a partner or small group

- Look at each object and complete.

corners _____

edges _____

faces _____

corners _____

edges _____

faces _____

corners _____

edges _____

faces _____

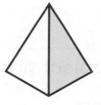

corners _____

edges _____

faces _____

- Discuss with your partner which 3D objects you think could be grouped together. List them here.

_____

Explain why you grouped these objects together.

_____

_____

Unit 17 **3D Objects** (TRB pp. 88–91)
Three-dimensional space MA1-14MG sorts, describes, represents and recognises familiar three-dimensional objects, including cones, cubes, cylinders, spheres and prisms

69

Guess My 3D Object

- Which 3D object has:

  * 8 corners

  * 12 edges

  * 6 faces that are rectangles? _____

- Draw two ways that this object is used in the school.

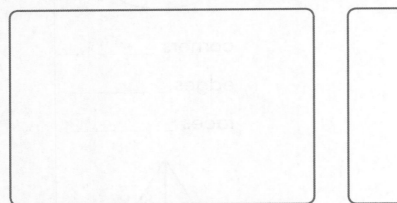

- Which 3D object has:

  * no corners

  * 2 edges

  * 2 faces that are circles? _____

- Draw two ways that this object is used in the school.

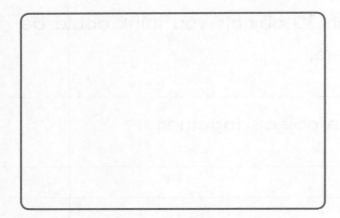
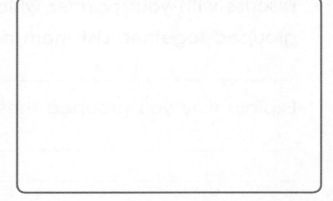

Unit 17 **3D Objects** (TRB pp. 88–91)
Three-dimensional space MA1-14MG sorts, describes, represents and recognises familiar three-dimensional objects, including cones, cubes, cylinders, spheres and prisms

DATE: _____

STUDENT ASSESSMENT

- Draw a line to match each 3D object with its name.

sphere   cube   cylinder   rectangular prism

  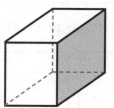 

- This is one **face** of a 3D object.

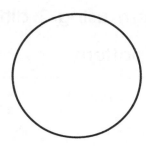

What 3D object could it be? _____

- Look at the 3D object.

  How many corners? _____

  How many edges? _____

  How many faces? _____

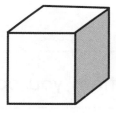

- A 3D object has:

  * 8 corners

  * 8 edges

  * 8 faces that are rectangles.

  Draw the 3D object.

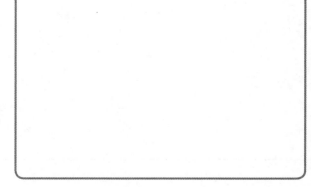

3D Objects (TRB pp. 88–91)
Three-dimensional space MA1-14MG sorts, describes, represents and recognises familiar three-dimensional objects, including cones, cubes, cylinders, spheres and prisms

71

Drawing Patterns

- Colour the shapes to make a pattern.

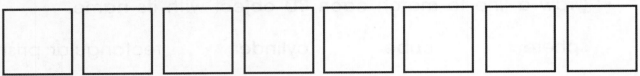

- Colour the shapes to make a different pattern.

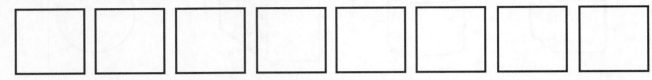

- Draw a pattern using 2 different shapes.

  Colour your pattern.

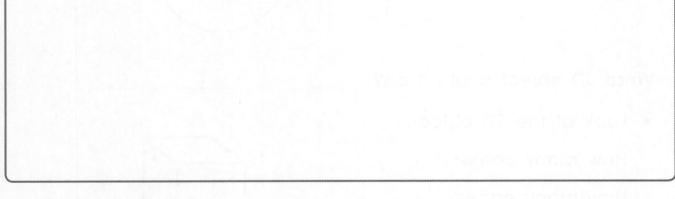

- Draw a pattern you can see in the classroom.

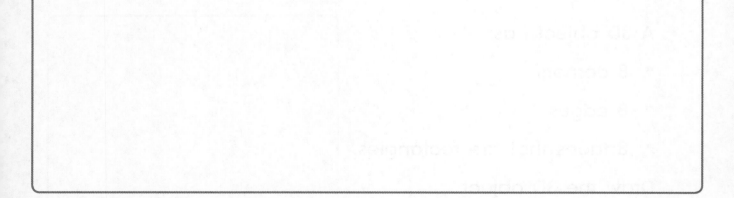

Describing Patterns

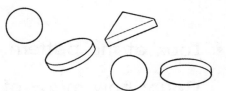

You will need: different shaped counters, a partner

- Trace over some counters to make a pattern. Colour the shapes in your pattern.

- Have your partner describe their pattern to you. Draw it below.

- Look at your partner's pattern and the pattern you drew for the second task. Are the patterns the same? Explain.

- Change one thing in your partner's pattern to make a new pattern. Draw it below.

Shape and Number Patterns

- Look at the pattern. Draw the next 4 shapes.

  Count how many of each shape and write the number pattern.

_____ _____ _____ _____ _____ _____

Look at the number pattern above. Draw another pattern to match.

- Make a shape pattern that matches the following number pattern.

  2   1   2   1   2   1   2   1   2   1

- Write the number pattern for the shape pattern.

- Draw another pattern to match the number pattern above.

STUDENT ASSESSMENT

• Copy the pattern.

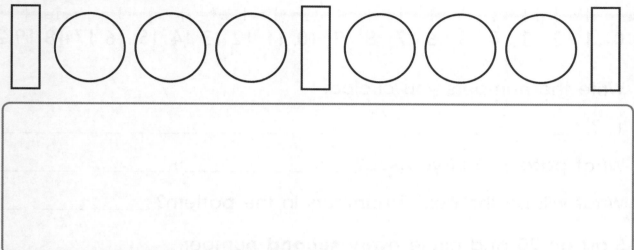

• Continue the pattern.

• Colour the shapes to make a pattern.

• Draw a shape pattern that matches the number pattern.

3	2	3	2	3	2

Unit
18
Patterns (TRB pp. 92–95)
Patterns and algebra MA1-8NA creates, represents and continues a variety of patterns with numbers and objects

75

Number-line Patterns

- Circle 1 and then every **second** number after that.

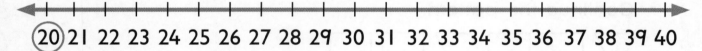

0 1 2 3 4 5 6 7 8 9 10 11 12 13 14 15 16 17 18 19 20

Write the numbers you circled.

1, 3, _____

What pattern can you see? _____

What will be the next 3 numbers in the pattern? _____, _____, _____

- Start on 20 and circle every **second** number.

20 21 22 23 24 25 26 27 28 29 30 31 32 33 34 35 36 37 38 39 40

Write the numbers you circled.

20, _____

What pattern can you see? _____

What will be the next 3 numbers in the pattern? _____, _____, _____

- Begin the number line at any number you choose.

Circle every **fifth** number.

Write the numbers you circled. _____

What pattern can you see? _____

What will be the next 3 numbers in the pattern? _____, _____, _____

Number Patterns (TRB pp. 96–99)
Patterns and algebra MA1-8NA creates, represents and continues a variety of patterns with numbers and objects

Colour the Numbers

- Colour 0 blue. Then count on 5 and colour that square blue, too. Continue colouring every fifth number blue.

0	1	2	3	4	5	6	7	8	9
10	11	12	13	14	15	16	17	18	19
20	21	22	23	24	25	26	27	28	29
30	31	32	33	34	35	36	37	38	39
40	41	42	43	44	45	46	47	48	49
50	51	52	53	54	55	56	57	58	59
60	61	62	63	64	65	66	67	68	69
70	71	72	73	74	75	76	77	78	79
80	81	82	83	84	85	86	87	88	89
90	91	92	93	94	95	96	97	98	99

Write the numbers you coloured.

_____

What pattern can you see? _____

- Look at the number chart. If you started counting from 1 and counted on 5, what would the first 10 numbers in the pattern be? List them below.

_____

- Look at the number chart. If you started counting from 8 and counted on 5 what would the first 10 numbers in the number pattern be? List them below.

_____

Unit 19   **Number Patterns** (TRB pp. 96–99)
Patterns and algebra MA1-8NA creates, represents and continues a variety of patterns with numbers and objects

77

Can You Find the Pattern?

You will need: a calculator, a partner

- Key your favourite number into the calculator.

  Then press + 5 = and keep pressing = .

  Write down your pattern.

  _____

- What is the final-digit pattern?

  _____

- Show your number pattern to your partner.

  Use the calculator to make your partner's pattern.

  Write down their pattern.

  _____

- Key in a number more than 50. Press − 2 = and keep pressing = .

  Record the number pattern you have made.

  _____

  What is the final-digit pattern?

  _____

  Is it the same as your partner? _____ Why?

  _____

STUDENT ASSESSMENT

DATE:

- Circle a starting number and **count on** to find a pattern that is counting by 5s.

5

0 1 2 3 4 5 6 7 8 9 10 11 12 13 14 15 16 17 18 19 20

Write the number pattern from the number line.

_____

What would the next 4 numbers be? _____

How did you work that out? _____

_____

- Finish the number patterns.

  6, 16, 26, 36, 46, _____, _____, _____

  3, 8, 13, 18, 23, 28, _____, _____, _____

  3, 5, 7, 9, 11, 13, 15, _____, _____, _____

- Finish the number patterns.

  48, 43, 38, 33, _____, _____, _____

  98, 88, 78, 68, _____, _____, _____

- Write a number pattern that ends with the following numbers

  _____1, _____6, _____1, _____6, _____1, _____6

Unit
19
Number Patterns (TRB pp. 96–99)
Patterns and algebra MA1-8NA creates, represents and continues a variety of patterns with numbers and objects

79

Money

- Draw two places that you might use money.

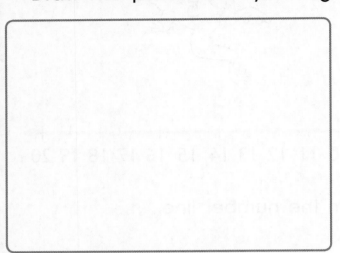

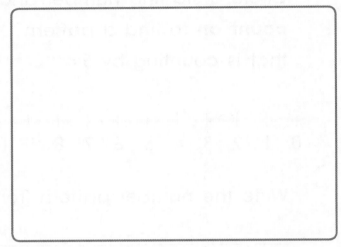

- Draw some money.

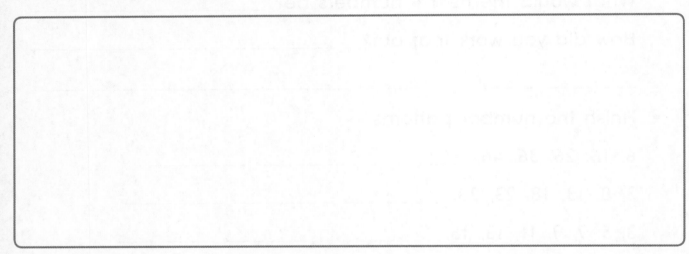

- If you had some money what would you do with it?

_____

_____

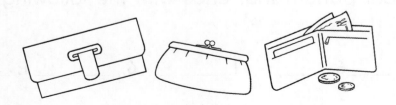

Coins, Coins, Coins!

You will need: copy of the top half of BLM 30 'Coins', scissors, glue

• Cut out the coins. Paste a coin in the correct box.

5c	20c
$1	$2
50c	10c

• Look at the coins. Write 5c, 10c, 20c or 50c under the coins.

_____   _____   _____   _____

Piggybank Game

You will need: a dice made from BLM 33 'Piggybank Dice', a partner

- In turn, roll the dice. Place a counter on the matching coin.

- If all those coins are covered, wait until your next turn.

- The first player to cover all the coins on the path wins.

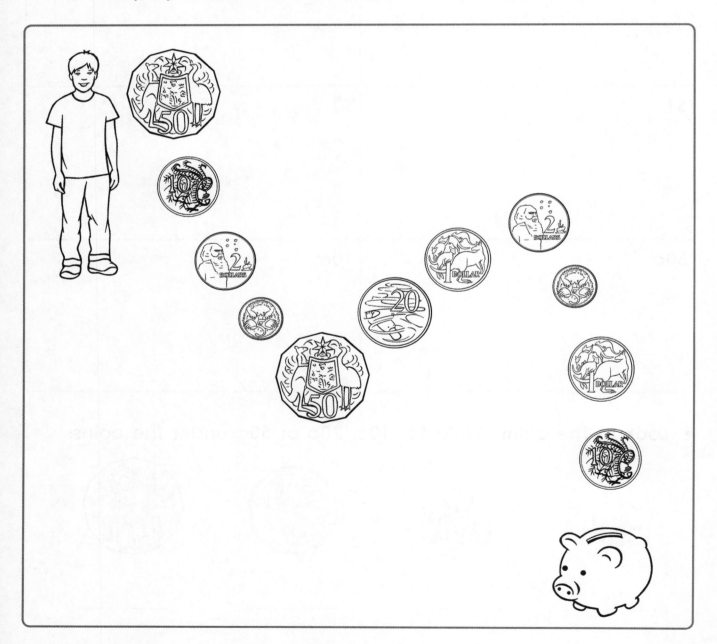

Unit 20 **Money** (TRB pp. 100–103)
Whole numbers (money) MA1-4NA applies place value, informally, to count, order, read and represent two- and three-digit numbers

20 STUDENT ASSESSMENT

DATE:

• Draw a place or situation where you might need to use money.

• Colour the coins.

• Draw lines matching the coins with their value.

| 5c | 10c | 20c | 50c |

Four in a Row

You will need: a partner, BLM 34 'Money Game Spinner', paperclip, sharp pencil

• Colour all the Australian coins.

• Play a game with a partner.

In turn, spin the spinner.

Circle a coin of that value.

The first player to circle 4 coins in a row wins.

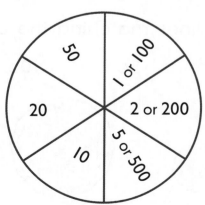

Coins in Order

You will need: 2 copies of BLM 30 'Coins', scissors, glue

• Cut out **one** of each coin. Paste the coins in order from the **smallest** coin to the **largest** coin. Write the value of each coin underneath.

• Cut out another set of coins. Paste them in order from the **lowest** value to the **highest** value. Write the value of each coin underneath.

• Draw some things you could buy for $1.

Unit 21 **More About Money** (TRB pp. 104–107)
Whole numbers (money) MA1-4NA applies place value, informally, to count, order, read and represent two- and three-digit numbers

85

How Much?

You will need: a copy of BLM 35 'Coin Bank'

- Use skip counting to work out how much money in each row.

Coins	Amount

- Draw some 5 cent coins that add together to make 30 cents. Or use the coins from BLM 35 'Coin Bank' and paste them below.

- Draw some other coins that add together to make 30 cents. Or use the coins from BLM 35 'Coin Bank' and paste them below.

DATE:

STUDENT ASSESSMENT

- Colour the Australian coins.

- Circle the coin with the **highest** value.

- Circle the coin with the **lowest** value.

- Write the value of **all** the coins.

      _____

- Write the value of **all** the coins.

      _____

- Write the value of **all** the coins.

     _____

In an Hour

 • Write the time. _____

• Show the time if one hour has gone by.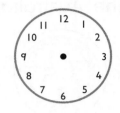

• Draw and write about something you might do **during** the one hour that went by.

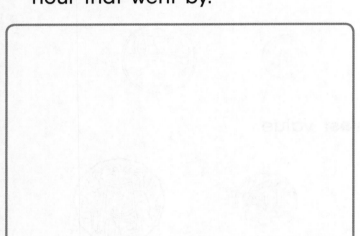

_____

_____

_____

_____

• Draw and write about something you might do that would take **less than** one hour.

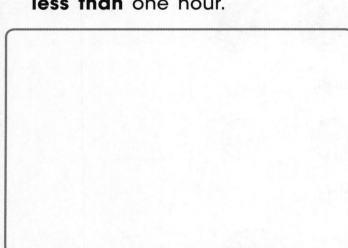

_____

_____

_____

_____

• Is recess longer or shorter than an hour? _____

School Week

- Draw your favourite thing to do on each day of the school week.

Monday	
Tuesday	
Wednesday	
Thursday	
Friday	

- Write your favourite day of the school week. _____

- How many days until your favourite day of the school week?

  _____

- How many days until the end of the school week? _____

- What day is it today? _____

- How many days until the same day next week? _____

Unit **22** **More About Time** (TRB pp. 108–111)
Time MA1-13MG describes, compares and orders durations of events, and reads half- and quarter-hour time

89

This Month

What month is it? _____

- Fill in the days of the week across the top of the calendar.

- Ask your teacher which day of the week the first day of the month is, and how many days there are in the month.

- Write in the days of the month.

- Circle today's date.

- How many days until the end of the month? _____

Monday						

Unit 22

STUDENT ASSESSMENT

DATE:

- Draw something you like to do that takes **more than** an hour.

- What day is it today? _____

- How many days until it is Monday? _____

- Look at the calendar below. How many days after the school excursion is Lee's birthday? _____

- Look at the calendar. How many days after Lee's birthday until the last day of the month? _____

October

Sunday	Monday	Tuesday	Wednesday	Thursday	Friday	Saturday
		1	2	3	4	5 Dad's Birthday
6	7	8	9	10	11	12
13	14 School Excursion	15	16	17	18	19
20	21	22	23 Lee's Birthday	24	25	26 Lee's Party
27	28	29	30	31 Halloween		

Unit 22

More About Time (TRB pp. 108–111)
Time MA1-13MG describes, compares and orders durations of events, and reads half- and quarter-hour time

91

Classroom Objects

- Look around your classroom. Draw 4 objects you think are heavy.

- What do you think is the **heaviest** object in the classroom?

  _____

- Pick up some light objects in the classroom.
  Draw 4 light objects.

- Of the light objects, which do you feel is the **heaviest**?

  _____

Mass (TRB pp. 112–115)
Mass MA1-12MG measures, records, compares and estimates the masses of objects using uniform informal units

Student Book

- Lift up your maths book and feel how heavy it is.

  Find 5 objects in your classroom that you think are **heavier than** your book.

  Heft your book and each object to make sure you are correct. Draw the objects below.

Heavier than

Of your 5 objects, which do you think is the **heaviest**?

_____

How did you decide? _____

- Find 5 objects that are **lighter than** your book. Check by hefting. Draw the objects below.

Lighter than

Of your 5 objects, which do you think is the **lightest**?

_____

How did you decide? _____

Unit 23 **Mass** (TRB pp. 112–115)
Mass MA1-12MG measures, records, compares and estimates the masses of objects using uniform informal units

93

Using Balance Scales

You will need: a pencil case, balance scales, classroom objects, a partner

- Work with a partner. Look around your classroom. You and your partner must find 2 objects each that you believe are **heavier than** your pencil case. Draw the 4 objects.

- Use the balance scales to check if you were correct. If your items were **heavier than** your pencil case, give them a tick.

- Now, you and your partner must find 2 objects each that you believe are **lighter than** your pencil case. Draw the 4 objects.

- Use the balance scales to check if you were correct. If your items were **lighter than** your pencil case, give them a tick.

DATE:

STUDENT ASSESSMENT

You will need: a copy of BLM 39 'Lots of Animals', scissors, glue, a shoe

- Cut out 2 animals and paste each animal in a box.

heavy	**light**

- Find something in the classroom that is **heavier than** your shoe. Draw it.

- Find something in the classroom that is **lighter than** your shoe. Draw it.

- A book is **heavier than** a pencil. Draw them on the balance scales.

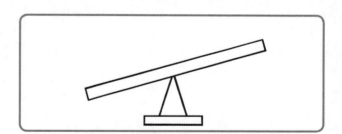

- An apple is **lighter than** a rock. Draw them on the balance scales.

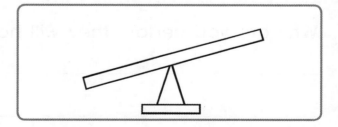

Unit 23 **Mass** (TRB pp. 112–115)
Mass MA1-12MG measures, records, compares and estimates the masses of objects using uniform informal units

95

Today at School

- Your teacher said that something was **impossible** to do at school today.

Draw what you think it could be.

Why would it be impossible?

_____

_____

_____

_____

Draw 2 things that you are **certain will** happen at school today.

Why are you certain they will happen?

_____

_____

Will it Happen or Not?

You will need: a copy of BLM 40 'What's the Chance?' between you and a partner, scissors, glue, a partner

- With your partner, cut out the pictures from BLM 40.

- Sort and paste the pictures into the things that 'will happen' at school today and the things that 'will not happen'.

will happen	**will not happen**

- Explain how you decided where to paste the pictures.

_____

_____

Tomorrow at School

- Label the pictures using the words below.

will happen   *will not happen*   *might happen*

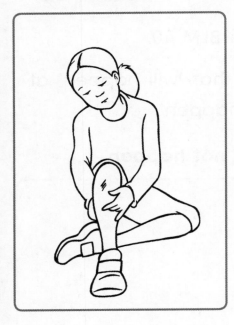

Draw your own pictures about school tomorrow.

will happen   **might happen**   **will not happen**

STUDENT ASSESSMENT

- Complete the sentence.

I am **certain** that today _____

_____

It is **impossible** for me to _____

_____

Tick the things that **will happen** today.

- Label the pictures using the words below.

will happen　　*will not happen*　　*might happen*

_____　　_____　　_____

Unit
24
Chance (TRB pp. 116–119)
Chance MA1-18SP recognises and describes the element of chance in everyday events

99

How Many Toys Left?

You will need: a dice

- Roll the dice. Write the number in the box. Cross out that many teddies. Now complete the number sentence.

8 – ☐ = _____

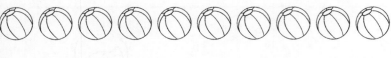

_____ – ☐ = _____

_____ – ☐ = _____

_____ – ☐ = _____

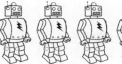

_____ – ☐ = _____

- Draw your own picture and write your own number sentence.

_____ – ☐ = _____

Subtraction (TRB pp. 120–123)
Addition and subtraction MA1-5NA uses a range of strategies and informal recording methods for addition and subtraction involving one- and two-digit numbers

Subtraction with Ten Frames

You will need: counters

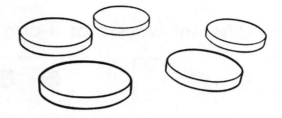

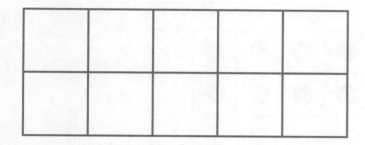 

• Use the ten frames and counters to solve these problems.

14 − 7 = _____   18 − 3 = _____

10 − 2 = _____   9 − 7 = _____

19 − 6 = _____   16 − 4 = _____

11 − 4 = _____   15 − 8 = _____

13 − 8 = _____   12 − 6 = _____

20 − 9 = _____   19 − 11 = _____

17 − 5 = _____   13 − 7 = _____

14 − 10 = _____   15 − 6 = _____

Unit 25 **Subtraction** (TRB pp. 120–123)
Addition and subtraction MA1-5NA uses a range of strategies and informal recording methods for addition and subtraction involving one- and two-digit numbers

101

Partitioning to Solve

- Look at the different ways that 8 can be **partitioned**.

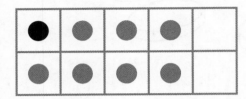   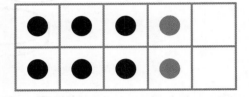

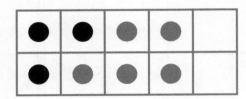   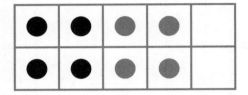

Use the ten frames above to help you solve the problems.

15 − 8 = _____   12 − 8 = _____   21 − 8 = _____

- Use the ten frames to show how 9 can be partitioned.
  Then solve the problems.

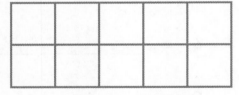   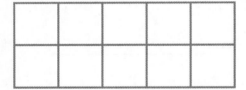

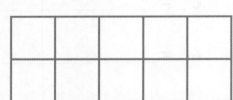   

17 − 9 = _____   22 − 9 = _____   13 − 9 = _____

- Solve the problems with what you know about 8 and 9.

17 − 8 = _____   15 − 9 = _____   20 − 8 = _____

15 − 9 = _____   14 − 8 = _____   16 − 9 = _____

STUDENT ASSESSMENT

- There were and 3 were taken away.

  How many were left?

  Write a number sentence for the problem.

  _____

- Use the ten frames to help you solve the problems.

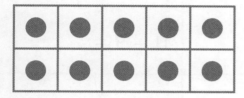

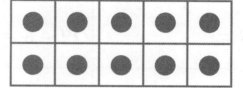

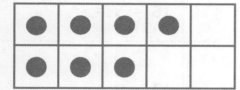

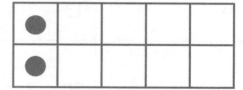

17 – 6 = _____   12 – 9 = _____

- Use the ten frames to help you solve the problems.

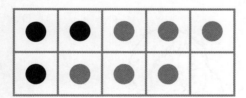

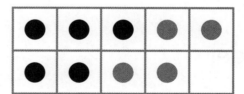

13 – 9 = _____   15 – 9 = _____

- What other problems could you solve?

  _____

  _____

Unit
25
Subtraction (TRB pp. 120–123)
Addition and subtraction MA1-5NA uses a range of strategies and informal recording methods for addition and subtraction involving one- and two-digit numbers

103

Solve with a Number Line

• Use the number line to help you solve the problems.

```
←+—+—+—+—+—+—+—+—+—+—+—+—+—+—+—+—+—+—+—+—+→
  0  1  2  3  4  5  6  7  8  9 10 11 12 13 14 15 16 17 18 19 20
```

20 − 7 =	15 − 8 =	18 − 4 =
13 − 9 =	19 − 6 =	11 − 9 =
15 − 3 =	13 − 7 =	16 − 5 =
17 − 4 =	20 − 5 =	12 − 8 =
19 − 11 =	14 − 9 =	17 − 8 =

Challenge!

The answer to a subtraction problem is 7. What could the problem have been?

More About Subtraction (TRB pp. 124–127)
Addition and subtraction MA1-5NA uses a range of strategies and informal recording methods for addition and subtraction involving one- and two-digit numbers

Three in a Row

You will need: cards made from a copy of BLM 44 'More Subtraction Problem Cards', a partner

```
8  –  6          15  –  7
```

- Select a card and work out the answer.

  In the grid below, colour the answer **red**.

  Your partner now selects a card and works out the answer. They colour the answer **blue**.

  Keep taking it in turn to select a card.

  The first player to colour 3 numbers in a row wins.

5	2	7	6
1	2	8	4
9	11	3	10
3	8	6	1

- Play again using your partner's book.

Unit **26** **More About Subtraction** (TRB pp. 124–127)
Addition and subtraction MA1-5NA uses a range of strategies and informal recording methods for addition and subtraction involving one- and two-digit numbers

105

Subtraction Story

- There were 12 apples on a tree.
  Draw the 12 apples.

- A man came and picked 5 of the apples.
  Cross out 5 apples.

  There were only _____ apples left.

  Complete the number sentence for the story.

  ☐ – ☐ = ☐

- Think of your own subtraction story.
  Draw a picture and write what happened.

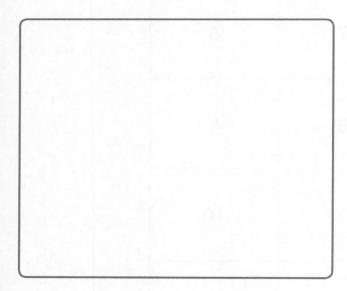

_____

_____

_____

_____

_____

Write the number sentence to match the story.

☐ – ☐ = ☐

More About Subtraction (TRB pp. 124–127)
Addition and subtraction MA1-5NA uses a range of strategies and informal recording methods for addition and subtraction involving one- and two-digit numbers

STUDENT ASSESSMENT

DATE:

- Use the number line to help you solve the problems.

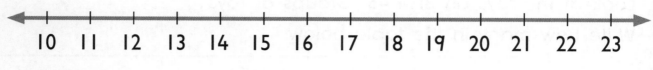

10 11 12 13 14 15 16 17 18 19 20 21 22 23

$16 - 5 =$ _____ $22 - 6 =$ _____ $15 - 3 =$ _____

- Count back to work out the problems. Circle the problems that equal 6.

$18 - 5 =$ _____ $13 - 7 =$ _____ $9 - 6 =$ _____

$11 - 5 =$ _____ $14 - 5 =$ _____ $15 - 4 =$ _____

- Solve this problem.

$17 - 5 =$ _____

- Draw a picture and write a number story.

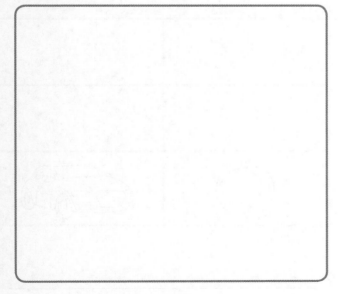

Unit
26
More About Subtraction (TRB pp. 124–127)
Addition and subtraction MA1-5NA uses a range of strategies and informal recording methods for addition and subtraction involving one- and two-digit numbers

107

Looking at Toys

You will need: a copy of BLM 45 'Groups of Toys'

- Look at the toys on BLM 45 'Groups of Toys'.
  Write how many in the table below.

- Use the data from the table to complete the graph below.

Title: _____

- Label your graph.
  What does your graph tell you? _____

_____

Favourite Toy

You will need: a partner

- Work with a partner. Choose 7 toys you think will be children's favourites. Write them below.

_____

- Write the 7 favourite toys in the table.

- Show your table to **10** people in your class. Ask: "What is your favourite toy?" Record their answers in the table. Don't forget to record your partner's favourite toy and your favourite toy, too!

Favourite Toys

- What is the favourite toy most children like? _____

- Why do you think it is the favourite? _____

_____

Unit 27 **Data** (TRB pp. 128–131)
Data MA1-17SP gathers and organises data, displays data in lists, tables and picture graphs, and interprets the results

109

Favourite Pet

You will need: a partner

- Ask 10 children:

  "Which animal would you most like to have as a pet?"

  Record their answers in the table.

- Look at your information. Colour the spaces to show how many children chose each animal.

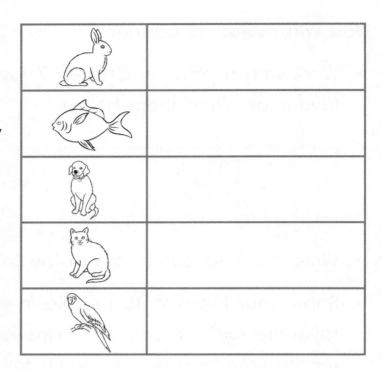

Favourite Pet

- What did you find out?

_____

STUDENT ASSESSMENT

- A class has drawn pictures of their pets. How would you group the information?

_____

- Look at the picture and fill in the table.

- Make a graph using the information in the table.

Title: _____

Unit
27
Data (TRB pp. 128–131)
Data MA1-17SP gathers and organises data, displays data in lists, tables and picture graphs, and interprets the results

111

What Number Could it Be?

- Put a ✗ where you think 18 would be on the number line.

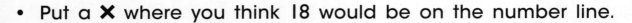

0   30

How did you decide? _____

- Put a ✗ where you think 35 would be on the number line.

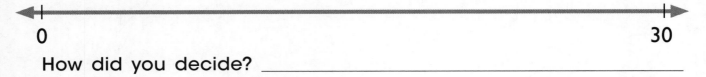

0   90

How did you decide? _____

- Put a ✗ where you think 9 would be on the number line.

0   50

How did you decide? _____

- Write what you think the number in the box could be.

0   30

How did you decide? _____

- Write what you think the number in the box could be.

0   70

How did you decide? _____

- Ali gave an answer 56. What do you think the question could be?

Place Value (TRB pp. 132–135)
Whole numbers MA1-4NA applies place value, informally, to count, order, read and represent two- and three-digit numbers

Number Bingo

You will need: cards made from BLM 47 'Place-value Bingo Cards' placed in a paper bag, a partner

- In the grid below, write any numbers more than 0 but less than 100.

- In turn, take out a card. If you have a number in your grid that matches the description on your card, then cross it off. You can only cross a number off once.

- Continue taking out cards. The first player to cross off all of their numbers wins.

Would you choose the same numbers again? _____
Explain.

_____

_____

Unit 28 **Place Value** (TRB pp. 132–135)
Whole numbers MA1-4NA applies place value, informally, to count, order, read and represent two- and three-digit numbers

113

Korean Finger Counting

- Look at the diagram.
  It will help you to show
  some numbers.

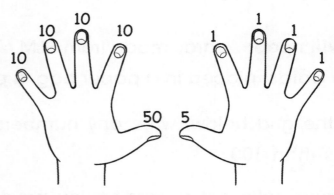

- Put crosses on the fingers you would use to show 8.

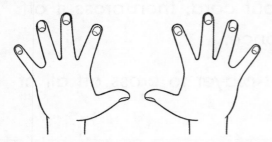

Here is the number
sentence that matches.

8 = 5 + 1 + 1 + 1

- Put crosses on the fingers you would use to show the numbers
  and write the number sentences.

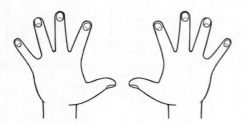
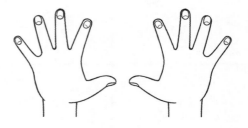

94 = _____   59 = _____

- List some numbers you could show by putting down 5 fingers
  on the table top.

STUDENT ASSESSMENT

• Write what you think the number in the box could be.

0   30

0   20

• What 2-digit numbers can you make from the following numbers?

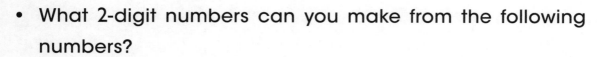

6   5   _____

Circle the **largest** number you made.

• Put crosses on the fingers you would use to show 28 using Korean finger counting.

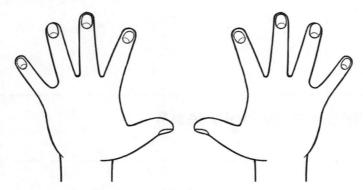

• Write the number sentence that shows the numbers you added together to make 28.

_____

• Use Korean finger counting to show 72.

Write the number sentence.

_____

Unit
28
Place Value (TRB pp. 132–135)
Whole numbers MA1-4NA applies place value, informally, to count, order, read and represent two- and three-digit numbers

115

Holds More, Less or the Same

You will need: a partner, a paper cup

- Find 5 containers in the classroom that **hold more** than a ⌄. Draw them below.

How did you choose your containers?

_____

_____

- Find 5 containers that **hold less** than a ⌄. Draw them below.

How do you know they hold less?

_____

_____

- Find a container that you think **holds the same** as a ⌄. Draw it below.

Capacity (TRB pp. 136–139)
Volume and capacity MA1-11MG measures, records, compares and estimates volumes and capacities using uniform informal units

Predicting

You will need: a partner, a paper cup, a small paper bowl, a spoon, some sand, rice or water

• Work with a partner. First **predict** how many spoonfuls of sand, rice or water each container will hold.

• Using the spoon, find out how much water, sand or rice each container holds.

I will use _____ to fill my container with.	Prediction	Actual measurement

• Explain how you made your predictions.

_____

_____

• Explain how you were sure that each container was full.

_____

_____

Unit 29 **Capacity** (TRB pp. 136–139)
Volume and capacity MA1-11MG measures, records, compares and estimates volumes and capacities using uniform informal units

117

Comparing

You will need: a partner; some containers; a spoon, scoop or small cup; some sand, rice or water

- Work with a partner. Choose 2 containers. Draw them in the table below.

- Circle the container you think will hold the **most**.

- Predict how many scoops, spoonfuls or cups of sand, rice or water the containers will hold.

- Measure to check your prediction.

  I will use a _____ *of* _____

  to measure the **2** *containers.*

Draw the containers.	Prediction	Actual measurement

- Which container holds the most? _____

  Explain how you know this.

  _____

  _____

DATE:

STUDENT ASSESSMENT

You will need: a cup-cake holder, rice, teaspoons

- Colour the container that holds the **most**.

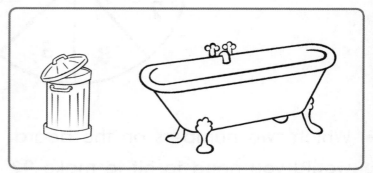

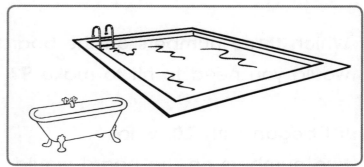

- How many teaspoons of rice do you think this  will hold?

- Use the teaspoon and rice to find out how much 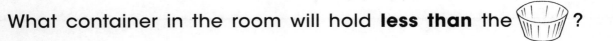 holds. How much does it hold?

- What container in the room will hold **less than** the ?

_____

- Explain how you know the container will hold **less**.

Unit
29
Capacity (TRB pp. 136–139)
Volume and capacity MA1-11MG measures, records, compares and estimates volumes and capacities using uniform informal units

119

Make 9

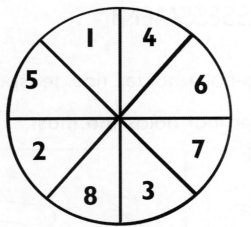

- Which **two** numbers on the board would you need to hit to make **9**?

- Which **three** numbers on the board would you need to hit to make **9**?

- If I began with **20**, what **two** numbers on the board would I take away to make **9**?

- If I began with **20**, what **three** numbers on the board would I take away to make **9**?

- Show some other ways you can make **9**.

Around We Go

- Complete the number trail. Begin at 7 and add 4. Write the answer in the next circle. Continue around the circle until you finish with the final answer 7.

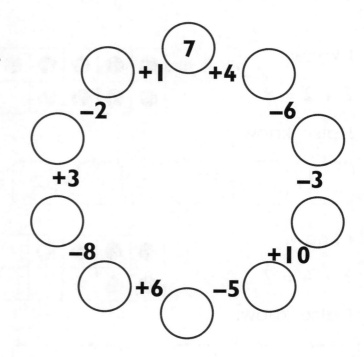

- Begin at 5. Think about what you need to do to get 9. Decide if you need to add (+) or take away (–). Write what you need to do on the line. Continue around the circle until you finish with the final answer 5.

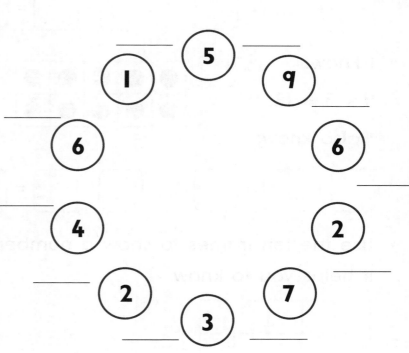

Explain how you knew to add or take away.

_____

_____

Unit **30** **Addition and Subtraction** (TRB pp. 140–143)
Addition and subtraction MA1-5NA uses a range of strategies and informal recording methods for addition and subtraction involving one- and two-digit numbers

121

I Also Know

- I know:

  $7 + 2 = 9$

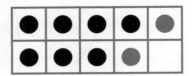

  I also know:

  ☐ + ☐ = ☐   ☐ − ☐ = ☐   ☐ − ☐ = ☐

- I know:

  $3 + 4 = 7$

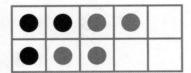

  I also know:

  ☐ + ☐ = ☐   ☐ − ☐ = ☐   ☐ − ☐ = ☐

- I know:

  $9 + 3 = 12$

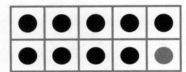 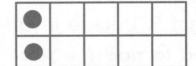

  I also know:

  ☐ + ☐ = ☐   ☐ − ☐ = ☐   ☐ − ☐ = ☐

- Use the ten frames to show a number fact and some other facts it helps you to know.

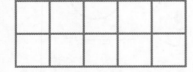

  ☐ + ☐ = ☐

  ☐ − ☐ = ☐

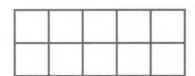

  ☐ + ☐ = ☐

  ☐ − ☐ = ☐

 Addition and Subtraction (TRB pp. 140–143)
Addition and subtraction MA1-5NA uses a range of strategies and informal recording methods for addition and subtraction involving one- and two-digit numbers

STUDENT ASSESSMENT

The target is 13.

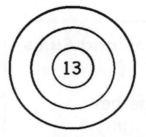

- Add 2 numbers to make 13.

- Add 3 numbers to make 13.

- Begin with 19 and make 13.

- Begin with 5 and make 13.

- Write the **hardest** way you can think of to make 13.

- If you know that 7 + 6 = 13, what else does this help you know?

Unit
30
Addition and Subtraction (TRB pp. 140–143)
Addition and subtraction MA1-5NA uses a range of strategies and informal recording methods for addition and subtraction involving one- and two-digit numbers

123

Maths Glossary

Mathematical Terms and Signs

+ **addition**
(and, altogether, combined with)

− **subtraction**
(take away)

× **multiplication**
(groups of)

÷ **division**
(shared between, how many groups of)

= **equals**

Fractions

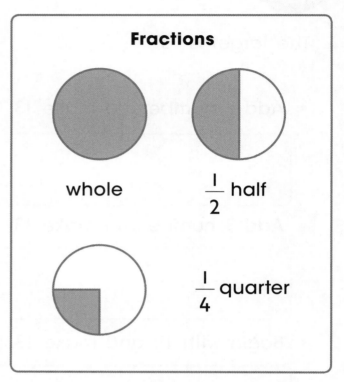

whole

$\frac{1}{2}$ half

$\frac{1}{4}$ quarter

2D Shapes

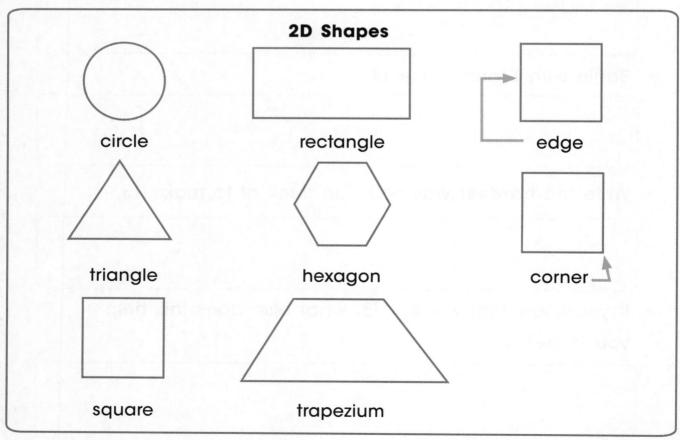

circle

rectangle

edge

triangle

hexagon

corner

square

trapezium

Maths Glossary

Time

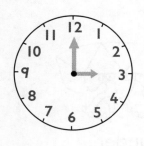

| 0 | 3 | : | 0 | 0 |

3 o'clock

| 0 | 9 | : | 3 | 0 |

half-past 9

Days of the Week

Sunday

Monday

Tuesday

Wednesday

Thursday

Friday

Saturday

Months of the Year

January

February

March

April

May

June

July

August

September

October

November

December

Numbers

0 zero	11 eleven	10 ten
1 one	12 twelve	20 twenty
2 two	13 thirteen	30 thirty
3 three	14 fourteen	40 forty
4 four	15 fifteen	50 fifty
5 five	16 sixteen	60 sixty
6 six	17 seventeen	70 seventy
7 seven	18 eighteen	80 eighty
8 eight	19 nineteen	90 ninety
9 nine	20 twenty	100 one hundred
10 ten		

Maths Glossary

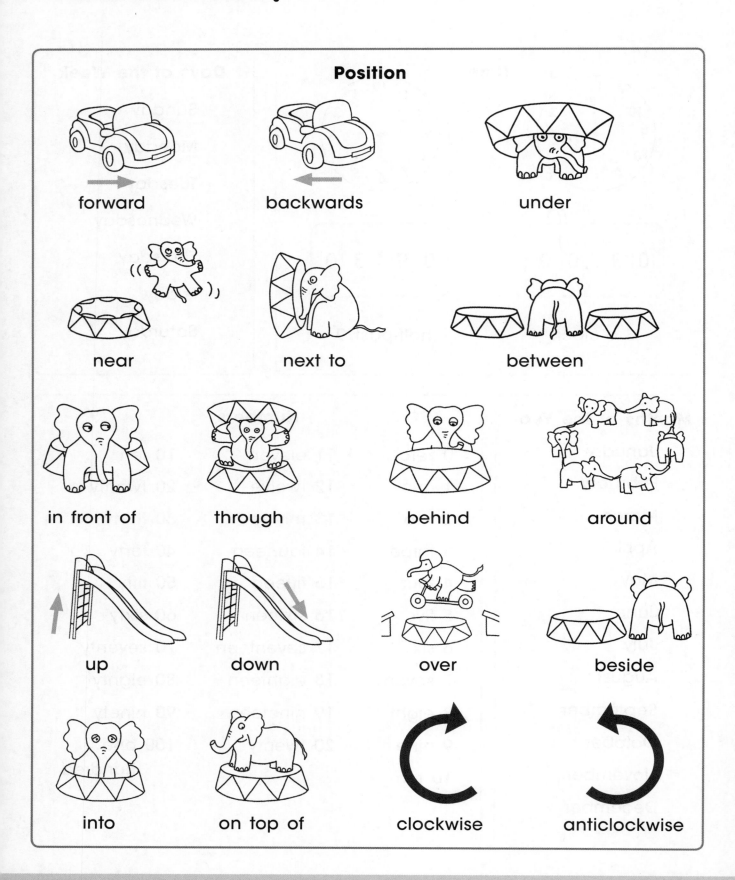

Position

forward

backwards

under

near

next to

between

in front of

through

behind

around

up

down

over

beside

into

on top of

clockwise

anticlockwise

Maths Glossary

Length

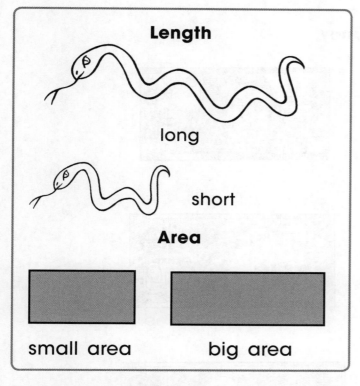

long

short

Area

small area   big area

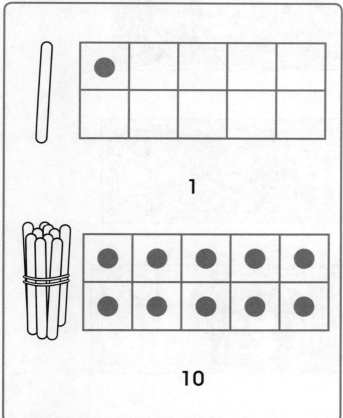

1

10

Mass

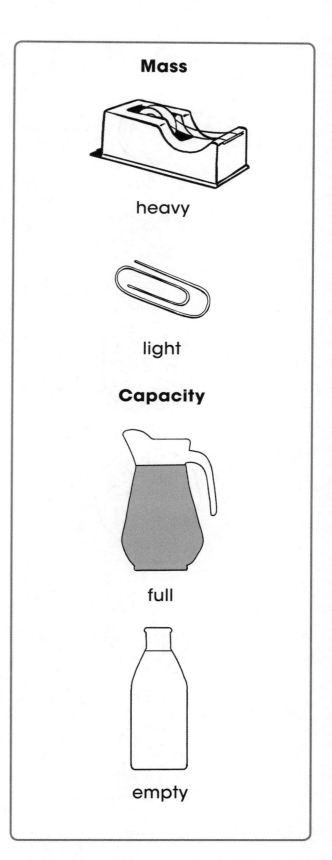

heavy

light

Capacity

full

empty

Maths Glossary

Money

5c

10c

20c

50c

$1

$2